Pittsburgh Antique Radio Society Publications
David W. Kraeuter, General Editor

1. *A Bibliography of Frank Conrad,* (Third Edition), 2007.
2. *The U.S. Patents of Reginald A. Fessenden,* (Second Edition), 2007.
3. *A New Bibliography of Reginald A. Fessenden,* (Second Edition), 2007.
4. *The U.S. Patents of John H. Hammond, Jr.,* (Second Edition), 2007.
5. *An Interview with Harold Beverage,* Richard Brewster, (Second Edition), 2007.
6. *Radio and Television Reminiscences: Raymond M. Bell in the* Pittsburgh Oscillator, (Third Edition), 2007.
7. *Electronic Essays*, David W. Kraeuter, (Fifth Edition), 2007.
8. *The U.S. Patents of Harold S. Black, Jack S. Kilby and Robert N. Noyce*, (Second Edition), 2007.
9. *The U.S. Patents of Stuart W. Seeley* (with a bibliography of Seeley's writings), (Second Edition), 2007.
10. *Ten Patents from Radio History*, David W. Kraeuter, 2007.
11. *Frank Conrad's Radio Patents: The Complete Texts*, (Second Edition), 2007.
12. *Electronic Reviews: Hundreds of Thoughts on 100 Books,* David W. Kraeuter, 2008.
13. *The 3 Strikes Camp Stories,* Karl Laurin, 2008.
14. *Vintage Radio Redux,* Karl Laurin, 2008.
15. *A Radio Patent Chronology*, David W. Kraeuter, 2009.
16. *25 Years of Electronic Reviews,* David W. Kraeuter, 2011.
17. *The Pittsburgh Antique Radio Society at 25*, 2011.
18. *Sketches from a Life in Electronics,* Laten Fetters, 2012.
19. *Mr. Johnson Answers*, William (Bill) Johnson, Jr. 2012.
20. *An Oscillator Reader*, 2013.
21. *Meetings, Contests, Winners*, 2013.
22. *An Oscillator Reader II*, 2014.
23. *The Pittsburgh Antique Radio Society at 30*, 2016.
24. *1 2 3 4 5 6: A David Kraeuter Sampler*, 2017.

Copies available at *www.lulu.com*

The Pittsburgh Antique Radio Society at 30

Commemorative

David W. Kraeuter, editor

Pittsburgh Antique Radio Society Publication 23

www.lulu.com

Presidents
Richard Brewster, 1986 - 1990
Richard Dreyfuss, 1990 - 1991
Richard Brewster, 1991 - 1994
Bonnie Novak, 1994 - 1998
Sev Dvorsky, 1998 - 2000
Bonnie Novak, 2000 - 2004
Regis Flaherty, 2004 - 2008
Chris Wells, 2008 - 2017
Don Merz, 2017 -

Honorary Members
Joseph Baudino, Raymond M. Bell, Richard Brewster,
Tom Dixon, Sev Dvorsky, G. Ray Fitterer, John Haught,
David Kraeuter, Bonnie Novak, Charles A. Ruch,
Alice Sapienza-Donnelly, Seth Ward.

Richard Dreyfuss, 1921 – 1991
PARS second President, 1990 – 1991

PARS pre-history
May 10, 1980; 454 Diablo Drive, Pittsburgh, Pa.
Seated, left to right, John Haught, Ted Core.
Standing, ?, ?, Jim Sutherland, ?
Help us identify these pre-historics.
(Photo: Richard Brewster).

1315 Gringo Road, Hopewell Township. PARS first meeting was at the Peoples Natural Gas Company, April 6, 1986. Anyone having photos taken at this meeting, please contact any PARS officer or Board member. (Photo: John W. Haught)

1986

The Pittsburgh Antique Radio Society holds its first meeting in Hopewell Township, Pa. Twenty-one people attend. Richard Brewster is elected our first president, April 6.

Our constitution is ratified. Article II : "The Society is dedicated to the preservation and exhibition of historic communications equipment and early electronic entertainment media, with an emphasis on the Pittsburgh area and related material."

Radio historian Raymond M. Bell addresses the society and becomes PARS first honorary member, and our first annual auction is held, September 13.

Dr. Raymond Bell (left) in discussion with David Kraeuter at September meet. (Photo: Ray Hill)

PARS holds a contest to determine a logo for the Club. The board chose Ray Hill's entry and the membership approved the choice. (Photo: Ray Hill)

Other logo contest entries.

**Bidding at September auction. This group was ready to buy.
(Photo: Ray Hill)**

**Rick Harris, Jr. (left), Laten Fetters and President Richard
Brewster at September meet. (Photo: Ray Hill)**

1987

Entries in our first contest, March 21. (Photo: Ray Hill)

We meet at the former home of radio pioneer Frank Conrad in Wilkinsburg, Pa., June 20. More than 100 people attend. Appropriately KDKA News filmed a feature story at the meeting and showed the film on KDKA TV news that night.

1988

Our first financial review reports an income of $1410.47 for 1987.

June meeting held at Bill Dawson's country home near Washington, Pa.

Ray Hill's license plate shows he obviously had the spirit. (Photo: Ray Hill)

1989

The first of many meetings at historic KDKA Hill Station in Forest Hills, March 18.

Westinghouse historian and sometime George Westinghouse impersonator Charles Ruch addresses the Society at the Westinghouse Lodge, March 18.

Bob Trauterman wins first John W. Haught Award for his display of Westinghouse items, March.

1990

Radio historian Ray Fitterer addresses the Society on Reginald Fessenden, March 17. Fitterer is elected a PARS honorary member.

President Richard Brewster (left) and Fessenden biographer Ray Fitterer at March meet. (Photo: Ray Hill)

Radio personalities Wendy King of KDKA radio talk show "Party Line," Bill Beal and Bill Hinds address the Society, June.

Bill Johnson, Jr.'s nostalgic take on Pittsburgh and radio history appears in the December *Oscillator*.

Frank Conrad historical marker dedicated outside his former home in Wilkinsburg, Pa., December 1. (Photo: DWK)

1991

PARS celebrates its fifth anniversary, March.

Membership soars to 104, June.

Founding PARS member Tom Dixon tunes in his Radiola 18 at our June 8 meet. (Photo: DWK)

We meet at historic educational television station WQED in Pittsburgh, December.

1992

Craig Dawson took first prize for his operating Stewart Warner 9100-B projection television in the March 21 contest. Note that the picture tube is mounted vertically, and the image is viewed via a mirror in the lid of the set. (Photo: Ray Hill)

PARS hosts the annual convention of the Antique Radio Club of America, Pittsburgh Airport Holiday Inn, June 4-6, 1992. Convention Coordinator: John W. Haught. (Photo: Ray Hill)

In a rousing speech at the Friday night, June 5 banquet, guest speaker Jim Conrad remembers his grandfather Frank Conrad, a KDKA founder. (Photo: Ray Hill)

Westinghouse Museum Tour and Dinner in the "Castle" in Wilmerding, Pa., December. Membership reaches 132.

1993

First "Spring Fever" meet is held in Washington, Pa., March 27. The event was sponsored by PARS, the West Virginia chapter of the Antique Radio Club of America (ARCA), the Mid-Atlantic Antique Radio club (MAARC) and the Museum of Radio and Technology.

PARS members Bob Rockwell, Jack Parsons, Bonnie Novak, John Haught, Seth Ward, Tom Dixon and Lee Schaeffer at June 19th meeting at Westinghouse Lodge. (Photo: Ray Hill)

Year-end banquet in Pittsburgh's Gateway Building 3, December 4.

Mark Cochenour's collection of two-transistor Japanese "boy's radios" from the '50s and '60s featured in the *Pittsburgh Oscillator,* December.

1994

Mike Adams's film *Broadcasting's Forgotten Father* shown at December meet. The film presents the story of forgotten radio pioneer Charles Herrold, who was broadcasting in California as early as 1909.

1995

Laten Fetters wins ugly radio contest, June.

Our tenth annual auction, September 16.

Authors: PARS honorary member Alice Sapienza-Donnelly and Rick Harris, Jr. ready to autograph their book at the December 2 meet. Bill Beal also assisted in writing *When Radio Was Young: Questions and Answers About Early Pittsburgh Radio.*

1996

Theodore Depto, P. E., becomes editor of *The Pittsburgh Oscillator; Journal of the Pittsburgh Antique Radio Society,* December. His tenure in that position was to last for the next 15 years.

1997

Seth Ward's obituary of Bruce Kelley, one of the founders of the Antique Wireless Association, appears in the December *Oscillator*.

Ted Depto's article on the unusual Montgomery Wards' Airline movie dial radio shares the same issue.

1998

Our 50th meeting, April 25.

1999

Westinghouse Lodge (Hill Station) deeded to Forest Hills Borough, January 20.

Death of honorary member Dr. Raymond M. Bell, radio historian who had contributed many articles to the *Oscillator*, April 12.

Westinghouse historian Ted Cook addresses the Society, December.

2000

April 8:"A Century of Recorded Sound"–one representative piece of recording equipment for each decade–1890 to 2000.

2001

Do It Yourself "Power Unit for Old Battery Sets" is published in September and December *Oscillator*. (See page 22.)

KDKA's Dave Crawley addresses PARS, December.

2002

Tenth "Spring Fever" meet held at Washington County Fairgrounds, March.

Karl Laurin's cartoon plays on the hyper-enthusiasm evident at many antique radio meets. See anyone you know here?

2003

Ernie Borelli shows his documentation on a transistor inventor at our June 14 meeting. (Photo: Andy Holovanisin)

Laten Fetters shows winning ribbon in the clock radio contest. (Photos from here through page 35 by DWK unless noted.)

Index to the first ten years of the *Oscillator* is published, September.

2004

PARS financial report for 2003 shows a $4,104.12 balance at the end of 2003. Total income for 2003: $7,178.85.

Crosley Pup and "Eighty PARS Meetings" featured in June *Oscillator*.

September *Oscillator* lists six PARS honorary members: Raymond M. Bell, G. Ray Fitterer, Joseph E. Baudino, Charles A. Ruch, Alice Sapienza-Donnelly and Seth Ward.

2005

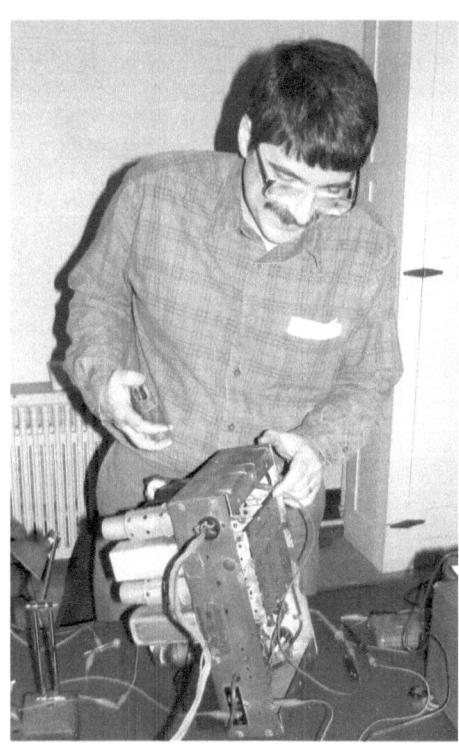

You break 'em, we fix 'em–PARS first workshop and service clinic boasted free service work on ailing antique radios, March 5. Louis Gaetano is seen here exercising his diagnostic skills. The program was so successful it has been repeated often.

First "Tri-state Radio Meet" at Holiday Inn, Beaver Falls, April 2.

Allen Benson's film "Tube" shown at our December meet.

2006

George Mitnick and Andy Dunlap take a break at our March 11 meet, which featured tombstone radios.

Our June contest was for sets with one to three tubes. Chris Wells is shown here with his prize-winning Westinghouse RC. Also taking home prizes that day were David Kraeuter for a Crosley Pup, and Craig Dawson for a home-brew set.

Rege Flaherty's beautifully-restored Philco 90 took first prize at our September 23 contest for eight or more tube sets.

Tim and Tom Duvall's rare Adams Paragon RF-5 and matching amplifier are shown at December meet. Karl Laurin restored the radio to operation, and explained its complicated tuning sequence.

Craig Dawson's crystal set of unknown make took first prize in December's contest.

2007

Tom Dixon's one-inch oscilloscope got a lot of attention at the June meet. The working RCA model 151 displayed the audio signal from a radio.

It isn't often that one sees the original shipping box for a 1937 console radio, but Chris Wells brought his to our June 23 contest featuring consoles.

A picture of the demolishing of the historic Westinghouse K building appears in September *Oscillator*. The roof of this building was the location of KDKA's famous first broadcast in November,1920.

PARS publications become available for sale on the internet through *www.lulu.com*, December. The series is growing on an irregular basis.

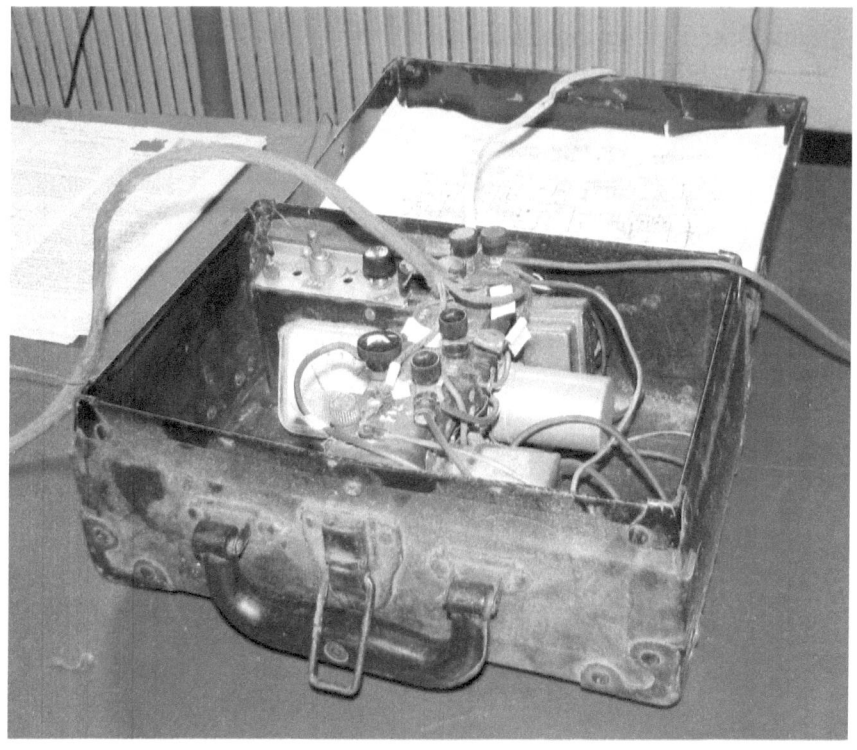

Shown at our March meeting, Karl Laurin's make-do home-brew portable power supply for battery sets is a classic. Articles describing construction of the apparatus appeared in the September and December 2001 *Oscillator*.

Our 101st meeting features a vintage tube tester contest, March 1.

Tri-state Radio Fest moves to Center Stage Banquet Hall in Monaca, Pa.

PARS publishes Karl Laurin's *Vintage Radio Redux*, a memoir of a life in radio servicing and tech school teaching.

PARS annual fall auction generates an impressive $2,600.

2009

Pittsburgh Oscillator editor Ted Depto at the April Radio Fest. Having edited the *Oscillator* for 15 years (approxlmately 800 pages), how can this man still be smiling?

Sev and Caroline Dvorsky at the April Radio Fest. Severin served as president of the Society from 1998 to 2000.

23

PARS "Old Radio on New Radio," a series of radio history essays, is broadcast on college FM radio station WNJR, Washington, Pa..

A smiling Sam Polis wants to talk about radios at our September meeting.

A smiling Lee Kronenberger wants to sell radios at our September meeting.

Efforts toward founding the National Museum of Broadcasting described in the December *Oscillator*. Priorities for the organization are reconstruction of the Frank Conrad garage and the establishment of a broadcasting history museum in southwestern Pennsylvania.

2010

The Society kicks off the new year scheduling an unprecedented eight meetings for 2010. Events included two radio service clinics, the sixth annual Tri-state Radio Fest, five contests, our equipment auction and our December luncheon.

Rege Flaherty at September meet (left); Dave Dellarosa speaks while Ed Baugher looks on at November meet.

Ken Young and John Malicky, December 4.

2011

(Facing page) Celebrating our 25th anniversary a little early at the March 29, 2011 PARS meeting are (from left): Bob Clark, Mike Ward, Andy Manko, Eugene Costel, Severin Dvorsky, Tom Dixon, Karl Laurin, Chris Wells, Charles Dudas, Ed Baugher, Patrick Pesce, Ed McGuigan, Dave Kraeuter, Joe Patrick, Jerry Sankovich, Andy Dunlap, Don Polito, George Mitnick and Charles Gessner.

Joe Patrick becomes the *Oscillator's* third editor, June/August.

Mike Elders donates two tablefuls of vintage radios and parts at the December 3 meet, all free to any members who wanted it—just the kind of spirit we like to see !

This gorgeous control escutcheon is part of Greg Bogel's rare Kelsworth chair side radio. On display at our June 25 meet. (Photo: Joe Patrick)

The Old Guard

Attendees at our April 10, 2011 meet who also attended our first meet on April 6, 1986 are:

Above, from left, Dave Kraeuter, Sev Dvorsky, John Haught and Tom Dixon. At left, Bill Hope.

Who will attend our 50th anniversary meeting?

2012

Mike Starcher's five-page article on building a one-transistor radio appeared in the March *Oscillator*.

Laten Fetters' rare Westinghouse SE1414 World War I aircraft radio on display at our June 23 meet.

Edward McGuigan began his coverage of the program side of radio history—as distinct from the hardware side—in the September *Oscillator*. He continued the subject with the presentation at our December meet of his "History of Sound Effects."

In his November 3rd presentation, Mark Hepburn showed us how useful the internet could be throughout a restoration project, in this case a Silvertone radio/phonograph combination.

Lou Gaetano showed us in great detail how to repair intermediate frequency transformers, in the December *Oscillator*.

2013

Chris Wells and Edward Schwartz presented the career of RCA radio engineer Raymond Berge and his involvement with the CRM-R6A and AR-8516 radiomarine receivers in the March *Oscillator*.

The words "charming", "rare" and "unusual" applied equally to Steve Geary's 1928 Zenith model 37A, first-place winner in this year's Tri-state Radio Fest.

On July 28 we revived the long-dormant idea of having a picnic at a PARS meet. The event took place at the Roosevelt Pavilion in North Park.

In one of the most unusual articles to appear in the *Oscillator,* Editor Joe Patrick and Raf Rincon gingerly investigated the delicate balance that can exist within an antique radio enthusiast's marriage.

Jeff Pleva showed us why we should stop passing by all those table clock radios from the 40's, 50's and 60's in his December article.

2014

PARS locates rare audio recording made by Frank Conrad and others. The recording was used by KDKA in a November 1941 broadcast in celebration of its 20th anniversary. See the March *Oscillator,* front page.

About 40 people attended the second annual picnic at North Park, July 27.

Extensive article by Mark Hepburn in the September *Oscillator* chronicles the complete restoration of a Stromberg Carlson 101H table AM radio.

In the same issue, Ed McGuigan provides a "genealogy"of radio.

Jeff Pleva teaches some PARS members the meaning of the word *dashpot* in the December *Oscillator.*

PARS Publication number 21 shows the where and when of all 150 PARS meetings, 1986 through 2014. It also lists the who and what of all 100 PARS contests for the same period, and tabulates all PARS meeting places. Plans are to update the publication at the end of each year.

Frank Conrad historical marker is rededicated and reinstalled in Wilkinsburg, Pa. The dedication service is attended by two of Conrad's great grandsons, October 17. (See page 10).

Frank Conrad's great grandsons Jamie Conrad (left) and David Conrad.

Joe Patrick's Silvertone display at December meet.

2015

At the March 28 meet, Jeff Pleva took us through the restoration of an RCA 45 rpm player.

May 3, 2015–John Marinacci's Atwater Kent model 5 takes first prize at the Tri-state Radio Fest. John then outdid himself by displaying his AK model 2. What an embarrassment of riches, John !

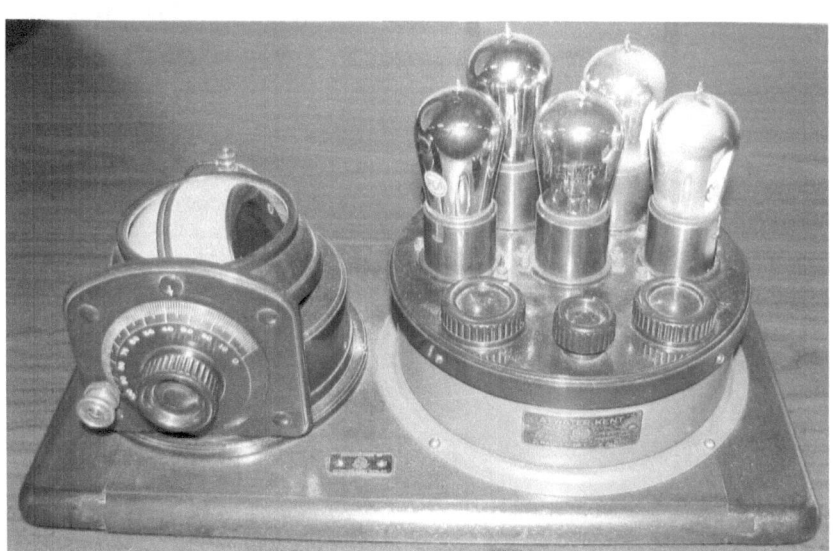

John's Atwater Kent model 5.

Grass roots volunteerism: Organizations that helped make the 11th Tri-state Radio Fest a success were: 90.1 FM WIUP, Just Radios, Universal Radio Inc., *The Latrobe Bulletin*, 1480 AM WCNS, Brett's Old Radios, Timesonline, Estes Auctions, Tube City Online, and 810 AM WEDO.

Lou Gaetano showed us how to build solid-state replacements for 6-volt vibrators in the March *Oscillator*. The replacements even fit inside the original vibrator cans. (But do they buzz, Lou?) Later in 2015 Lou followed up with articles about re-stuffing Philco bakelite capacitors and building a one-tube AM radio using tubes originally designed for use in TVs. Just the kind of Do It Yourself stuff we like to see.

Heathkit is back–again (maybe). The December *Oscillator* front page features Heath's new *TRF* AM receiver kit, the GR-150 at $149.50. See more at: *http://shop.heathkit.com.*

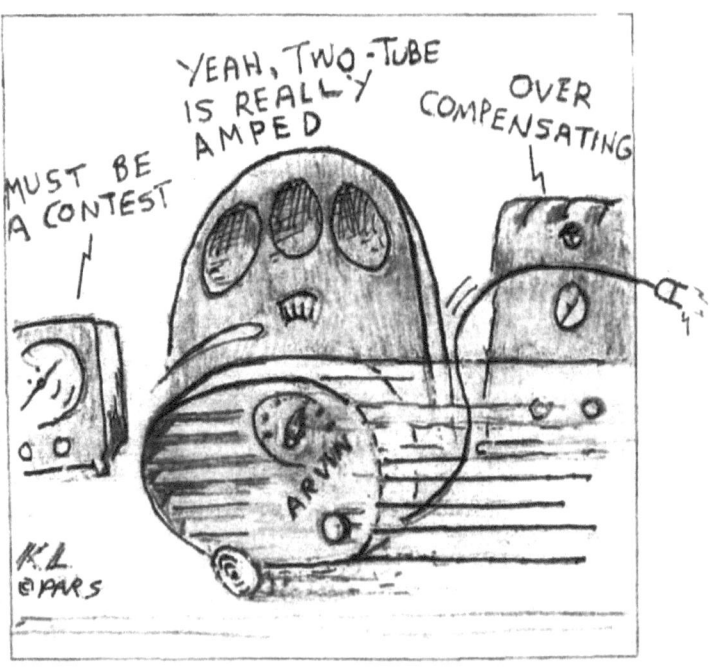

From the June *Oscillator*–the inimitable Laurin.

2016

Regis Flaherty
February 6, 1952 – January 15, 2016
PARS President, 2004 – 2008

Rege conducting the September 23, 2006 PARS meeting.

(Facing page) Celebrating our 30th anniversary a little early at the March 12, 2016 PARS meeting are (standing, from left): John Bruno, Mike Elders, Richard Constantino, Don Polito, Andy Dunlap, Karl Laurin, Tom Rohac, Don Merz, Tim Tress, Ed McGuigan, Louis Gaetano, Chris Wells, Jeff Pleva, Harry Brathoover, Kevin Gontz, Rafael Rincon, Tom Dixon, Sev Dvorsky, Andy Manko, Doug Ramsay, and George Mitnick.

(Second row, from left): Mike Sapp, Andy Holovanisin, Joe Patrick, Dennis Suhre, Dave Kraeuter, Jonathan Brady, Anders Hammer. (Photo: Mary Brady)

A Few Final Thoughts for Historical Continuity.

Raymond Martin Bell (1907-1999), PARS first honorary member, could name the day–May 18, 1922–he first heard a radio. He is shown here in about 1925 with his Reinartz set, which he finished building on May 12, 1923. In his senior year in college in 1928 he built and demonstrated a closed-circuit mechanical tv, then wrote about that experience 71 years later in the March 1999 *Pittsburgh Oscillator*. (Photo courtesy of Marty Butler)

Schematic for Reinartz set.

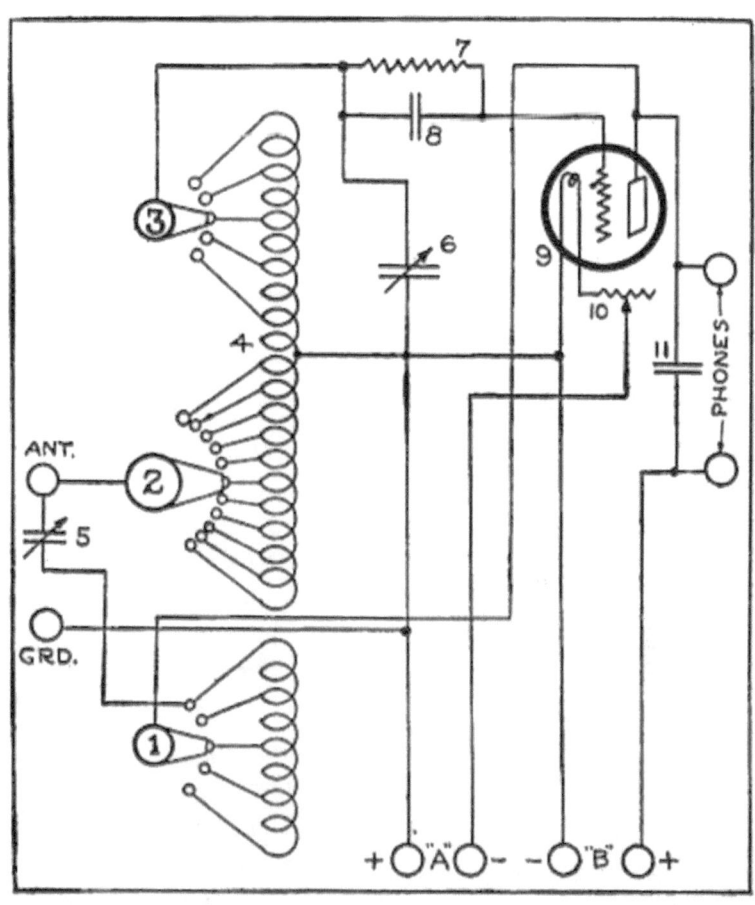

Radio Age, vol. three, no. 7, November 25, 1922.

Personal Name Index

Notes

www.ingramcontent.com/pod-product-compliance
Lightning Source LLC
Chambersburg PA
CBHW021940170526
45157CB00005B/2361